Heavy Sidecar Motorcycles of the Wehrmacht 1935-1945

Obviously, a cycle with such a wide fender was not designed for military use. The R 12 was also available on the civilian motorcycle market with gloss paint at that time. This BMW R 12 mounted rifle cycle came from the first series built, delivered to the newly formed Wehrmacht after 1935.

Horst Hinrichsen

Schiffer Military History
Atglen, PA

The motorcyclist's driving-school training did not include carrying out minor repairs to the vehicle while underway, but it was a very popular show to put on for the public.

Translated from the German by Ed Force

Printed in China.
ISBN: 0-7643-1272-3

This book was originally published under the title,
Waffen Arsenal-Schwere Beiwagenkräder der Wehrmacht 1935-1945
by Podzun-Pallas Verlag, GmbH

We are interested in hearing from authors with book ideas on related topics.

Published by Schiffer Publishing Ltd.
4880 Lower Valley Road
Atglen, PA 19310
Phone: (610) 593-1777
FAX: (610) 593-2002
E-mail: Schifferbk@aol.com.
Visit our web site at: www.schifferbooks.com
Please write for a free catalog.
This book may be purchased from the publisher.
Please include $3.95 postage.
Try your bookstore first.

In Europe, Schiffer books are distributed by:
Bushwood Books
6 Marksbury Ave.
Kew Gardens
Surrey TW9 4JF
England
Phone: 44 (0) 20 8392-8585
FAX: 44 (0) 20 8392-9876
E-mail: Bushwd@aol.com.
Free postage in the UK. Europe: air mail at cost.
Try your bookstore first.

Heavy Sidecar Motorcycles of the Wehrmacht 1935-1945
General Information

Heavy motorcycles with sidecars—"Beiwagenkräder" in military language—enjoyed great popularity among the Reichswehr and Wehrmacht soldiers from an early date.

This is understandable, for on the German civilian motorcycle market in the thirties, cycles with sidecars were heavily advertised. The economic rise after 1933, spurred by the treacherous hopes of the Third Reich, surely contributed to this. Although it was usually only the solo motorcyclist who could enjoy the fun of exciting overland trips until then, the motorcycle industry then began to produce many comfortable sidecars. Low-priced small cars, or three-wheelers, were also on the market in the thirties. But if a buyer decided in favor of a motorcycle with a sidecar, this decision could surely be attributed to that era's intensive advertising for sidecar rigs. Sidecar cycles with elegant sheet-steel box frames, Cardan drives, and four-speed transmissions were also a luxury for frustrated auto drivers.

Without a doubt, there were also previous designs of a sidecar type. For the most part, though, they did not live up to the requirements of letting usually feminine sidecar riders take part in a cycle ride without problems and with protection from wind and weather. Only the touring types with handy entrance steps, windshields, and swing axles offered the required comfort for the sidecar passenger. Some manufacturers delivered complete sidecar rigs, while others produced only sidecars, which could then be attached to the motorcycles by their manufacturers. Such sidecar machines were introduced into the Reichswehr in fairly large numbers, either painted gray or in spotty camouflage, and later taken over by the Wehrmacht. But then, as previously, there were pure road machines, which had considerable problems in off-road use, with an average 10-12 cm of ground clearance under the cycle and about 25 cm under the sidecar.

Between 1929 and 1934, the Reichswehr took large numbers of BMW R 11 sidecar cycles into their service, to be taken over by the Wehrmacht in 1935. With stable box frames and powerful engines of 750 cc displacement, these cycles were particularly useful as heavy sidecar cycles.

Only the type of classification was clearly regulated in the Wehrmacht. Thus, the machines with displacements of more than 500 cc were classified as heavy sidecar cycles. Therefore, the Zündapp K 500, in a special version that was already used in prewar times as the so-called "Official Version," was listed in Vorschrift D 600 "Durability Values of Motor Vehicles and Equipment" of April 10, 1940, and was still classed as a medium cycle.

Since "Reichswehr cycles" were still widely used later in the Wehrmacht, they will also be mentioned here, even if they were first used before 1935.

Some of the models mentioned below were also used in solo form by the Reichswehr and Wehrmacht. These solo cycles will not be noted here.

At first, sidecar cycles were intended only as transport vehicles to bring infantrymen as close as possible to the front lines quickly. As a rule, they did not leave permanent roads and paths, since their optimal use as road machines was possible only there. The sidecar cycles available from civilian series production by the beginning of World War II were well suited to such use. But in the western campaign in the roadless fields of Flanders, these road machines already had serious problems—to say nothing of the eastern campaign.

Only the off-road machines designed much later, the BMW R 75 and Zündapp KS 750 of the motorcycle riflemen, a troop type already introduced as of 1935, allowed the use of these overly heavy sidecar cycles in off-road areas in the middle of World War II. Through greater ground clearance, especially large off-road profiled tires, road and off-road gears, reverse gear, and many other technical innovations, these cycles first corresponded to the requirements of an off-road-capable vehicle for motorcycle riflemen. But, by the time these cycles were actually available to the troops, it was often too late. The cycle rifle troops had already been practically wiped out in hard fighting by this time. So the cycles were then made available for the most part to the remaining cycle riflemen, who now had to carry out reconnaissance and scouting tasks. Individual cycles were also turned over to the supply troops. Many a motorcyclist could hardly wait until his BMW R 75 or Zündapp KS 750 arrived at the front, to replace the series-production civilian machine he had been using. In the Russian winter war, in particular, the motorcyclists could really appreciate these cycles, as the R 75 and the KS 750 had not only sidecar drive, but also hand- and foot-warmers. The handgrips and footrests, heated by warm air, gave the driver a pleasant warmth and prevented his limbs from freezing. In fact, even the interior of the sidecar was warmed by a spiral-shaped heater, so that the possibility of riding in a sidecar of such a cycle in winter was much desired.

This BMW R 11 also ranked among the Wehrmacht's older sidecar cycles at the beginning of World War II as its external state shows. Yet it was obviously fully ready for service for the purpose of carrying supplies of wine.

Throughout the war, heavy sidecar cycles were thus used by all service arms for transport of such persons as paymasters, ostlers, and company or platoon leaders when they were called briefly to meetings or other events. They were also used by motorcycle messengers or couriers, although here the light solo cycles were used on principle.

The loss of sidecar cycles was already very high at the beginning of the war, so that the requisitioning of civilian machines, wisely begun even before the war broke out, paid off. Later on, motorcycles and their owners often went into service together. This resulted in an unmanageable variety of motorcycle types, which resulted in a constant shortage of spare parts that was a major problem in making repairs. Then, too, in occupied areas local workers were often used in repair shops to free German soldiers for military duties; thus, repair work was often delayed because of language problems. Surely these problems also affected other branches of the Wehrmacht. They, too, deserve mention within the parameters of this subject.

When we speak of the use of heavy sidecar motorcycles, we find two manufacturers, BMW and Zündapp, standing out, as their cycles were designed primarily for military use. But we can also name Victoria, NSU, and DKW. Their part in supplying heavy sidecar motorcycles was considerably smaller than that of BMW and Zündapp. Economic considerations surely played a role here, in that smaller manufacturers were also given state contracts.

INDIVIDUAL DETAILS

1. BMW R 11

The two-cylinder BMW R 11, of 750 cc displacement, was used by the Reichswehr between 1929 and 1934, and saw partial service until the end of World War II. This was the first BMW cycle with a stable box frame, which was well adapted to use as a sidecar cycle in daily military action. During its five years in production, it was constantly undergoing technical improvements. For example, the wide fenders originally used were replaced by narrower ones, so that much less mud could get stuck around the wheels during off-road use.

2. VICTORIA KR 6 BERGMEISTER

A similar fate was that of the Victoria KR 6 Bergmeister, which displaced 600 cc and was used almost always as a sidecar cycle in the Reichswehr from 1933 on, and was then taken over by the Wehrmacht—often as a driving-school motorcycle. Some 3,000 of these cycles were used in the Reichswehr. The successor models, KR 8 and KR 9, will be mentioned here only in passing, since only small numbers of them were used as sidecar cycles.

These two models were the most popular heavy sidecar cycles of the Reichswehr era that were still found in goodly numbers in the Wehrmacht. After the war broke out, they were used not only at army bases, but also appear at times in action photos from the front lines.

3. BMW R 12

The sidecar cycle most often found in the Wehrmacht was the BMW R 12. It was already introduced for troop use in 1935, and over 36,000 vehicles had been delivered to the

A motorcycle rifle batallion in field parade position on the occasion of a parade in Giebelstadt, near Würzburg, on September 17, 1936. Their uniform equipping with BMW R 11 cycles is impressive.

Wehrmacht by 1941. It became, after several modifications, the basic equipment for the motorcycle rifle troops that originated in 1935. With a 746 cc, 18 hp two-cylinder four-stroke engine, it was the first BMW model that featured a four-speed gearbox and a telescopic fork with hydraulic shock absorbers. The ground clearance under the vehicle was only 12 centimeters, so that in spite of everything, it was not a genuine off-road machine.

4. ZÜNDAPP K 800 W

Zündapp also participated in supplying the Army with sidecar cycles. The Wehrmacht cycle with the largest displacement was the Zündapp K 800, which came from civilian production; some 5,000 of them were delivered to the Wehrmacht by 1938. Equipped with saddlebags and painted in Wehrmacht gray, it was designated K 800 W. It was the only four-cylinder cycle among the Wehrmacht's sidecar motorcycles, and achieved success even before the war in many Wehrmacht off-road competitions. Like the KS 750 (see below), it remained in service to the war's end.

5. ZÜNDAPP KS 600 W

The KS 600 W was the successor to the K 800 W. Essentially, it was an "enlarged" K 500. Equipped with the special official sidecar of the Type 39, developed by Zündapp and the Stoye firm of Leipzig, some 18,000 of this model reached the Wehrmacht between 1938 and 1941. But this machine did not live up to the demands that were made by the expansion of combat activities in the east, even though the KS 600 W was conceived for military use. Thus, the delivery of this model decreased significantly to the end of 1941, since the Wehrmacht command came to realize that this model did not live up to the demands of the rough terrain. Thus, Zündapp decided to develop a completely new sidecar cycle made especially for off-road use.

6. BMW R 61

The BMW R 61 was also used by the Wehrmacht between 1938 and 1941, sometimes as a sidecar cycle. With 18 hp and 600 cc displacement, this sidecar cycle also ranked among the Wehrmacht's standard equipment. The cycle was the first Wehrmacht model to have rear axle suspension.

7. BMW R 71

Having greater engine power than the R 61, and thus better suited to sidecar use, the R 71 had 750 cc displacement and 22 hp. Otherwise, its technical details were the same as those of the R 61. This sidecar cycle was likewise used by the Wehrmacht from 1938 on.

8. ZÜNDAPP KS 750

As already noted, the Zündapp KS 750, along with the BMW R 75, was introduced into Wehrmacht service in 1941. Here again, it was a new design that had been developed since 1937. The KS 750, like the R 75, had surely been inspired by such heavy Belgian and French sidecar cycles as the FN M 12, Sarolea, Gillet 750, and Gnome et Rhone AX 2, which were already in military use by those two countries before the war began.

Motorcycle riflemen at rest. The unit was supplied with BMW R 11 rigs. Note the deflector bar in front of the cylinders, found only on the military versions, to protect the low-lying cylinders from damage in off-road use.

9. BMW R 75

At the same time as the Zündapp KS 750, the overly heavy BMW R 75 reached the troops in the middle of World War II. Like the Zündapp already named, it was intended to be a sidecar-driven cycle for use by the cycle riflemen. With a total of ten road and off-road gears, the driven sidecar wheel, and greater ground clearance than any other motorcycle, a cycle was available for the first time that lived up to the demands of an off-road-capable action motorcycle. With 750 cc displacement and 26 hp, its performance was better than the earlier BMW models. With a net weight of 420 kg, it was capable of carrying another 420 kg. A further innovation was a frame screwed together from individual parts. Thus, it was possible to repair damage by replacing individual segments of the frame without taking the entire cycle apart. The exhaust was mounted high, hand and foot heating could be installed for winter use, and the air filter was moved to above the fuel tank for use in North Africa. In all, the cycle seemed very massive, with the large wheels and rough-treaded tires adding to this impression.

10. OTHERS

Among the other heavy sidecar motorcycles that were taken from civilian production and used by the Wehrmacht with minor modifications, but which were not very widespread or often seen because of their small numbers, are the NSU 601 OSL and DKW 50C. It is certain that nothing but the Wehrmacht gray paint was their only military feature. From the occupied lands of France and Belgium there also came captured motorcycles that were utilized by the German troops to lessen the constant shortage of their own German cycles. Above all, there were the heavy sidecar rigs by FN, Gnome et Rhone, and Gilet that were used by the German Army. As a rule, they had no great "life span," for the German motorcyclists lacked the requisite knowledge of how to maintain those cycles. Only for the captured French Gnome et Rhone AX 2 was a service manual printed in German.

The use of captured cycles as heavy sidecar rigs surely made the problems of finding off-road cycles greater instead of smaller.

When they were in a hurry, the motorcyclists often gave their orders in a loud voice without coming to a stop, as this messenger on a BMW R 12 rig is doing during the French campaign.

The BMW R 12 sidecar cycle looks small and fragile beneath the wing of a Junker 88 on a French military airfield.

Among the youth of the prewar generation, military vehicles, like the BMW R 12 sidecar rig seen here, were well worth admiring, and their owners were proud to have pictures of them taken. Often the young folks' fathers, brothers, or other close relatives were in service.

Trips for further training in peacetime and in good weather have always been regarded by motorcyclists as one of their more pleasant military activities. Here BMW R 12 sidecar cycles are leading the column

Above: The B Motorcycle Squadron of Heavy Assault Gun Unit 197 awaits the command to march (Balkan campaign, 1941). In the center is, among others, a BMW R 12 sidecar cycle.

Here it was the medics who were issued a formerly civilian sidecar cycle as transport for an army doctor. This is obviously an NSU 601 OSL, with a classy sidecar featuring an attaching bar, a curved bumper, and a removable windshield.

The different fenders on these BMW R 12 cycles can be seen clearly. The sidecar cycle on the left has the first version of the front fender, while the one on the right already has a modified version.

The variety of different motorcycles used by the Wehrmacht was easy to see when German troops marched into the Sudetenland in 1938. In the foreground is a BMW R 12 sidecar cycle with a black factory paint job.

The crew of a sidecar cycle formed a devoted team. This was especially true of the motorcycle riflemen, like the three soldiers who stand by their BMW R 12 here.

This BMW R 12 was obviously used by the motorcycle riflemen, too. The swinging rack for the MG 34 is still on the sidecar. The crew's baggage is strapped onto the spare tire. Note also the horseshoe on the sidecar.

The pack train of an unknown unit at rest. The BMW R 12 sidecar cycle with the wide front fender is missing one headlight cover.

This crew from a motorcycle rifle unit carries a mount for an MG 34 on their BMW R 12 rig. The heavy machine gun (sMG) could also be set up on a tripod for use as an anti-aircraft gun. The sMG was the most vital weapon of the motorcycle riflemen.

Left: A motorcycle messenger of the 291st Infantry Division on a Zündapp KS 600 W sidecar rig. The KS 600, with its 28 hp, also ranked among the heavy sidecar cycles. It differed from the civilian road machine only by having gray paint, a masked headlight, and saddlebags.

Right: Members of the Guderian Panzer Group have settled in the basement garage of a French villa during the French campaign. In the foreground is a BMW R 12 with a wide fender from the first production run.

Below: On the sidecar of this heavy BMW R 12 rig, the MG 34 has been mounted on its bracket and wrapped to keep out dust and prevent damage during a march in enemy-free country.

This handsome civilian Zündapp K 800 with Steib No. 28a sport sidecar still has its gleaming original factory paint job...

...but soon this cycle, too, will get a coat of gray camouflage paint.

Left: This driver presents his BMW R 12 rig in prescribed gray paint. He wears the so-called motorcycle coat, which was introduced as a uniform piece for motorcycle drivers in 1934, along with the uniform goggles and gloves.

Right: Thus, a BMW R 12 sidecar cycle's career ended in the 1941 Russian campaign.

Below: While still on German soil—as here in the vicinity of Königsberg in 1940—the crews had a chance to test their cycles. Here a Zündapp K 800 W rig needs assistance from human muscle power.

Left: A quick look at the map was necessary when the crew of this Zündapp KS 750 reached the French-Spanish border.

Lower right: Even the robust BMW R 11 sidecar cycles with their 11 hp were not made for such terrain. Only the towing power of a tracked vehicle can get them out of this miserable situation.

Bottom: A motorcycle rifle platoon with uniform BMW R 11 cycles, seen in training in Austria on March 23, 1938. The sidecars still have doors, which slowed the riflemen's entrances and exits. Later, a cutout in the sidecar body replaced the door.

A Zündapp KS 600 W with a 34 light machine gun on the sidecar. Targets could be taken under fire when stopped, and often while in motion, too, though without aiming.

Left: The successor to the K 800 W, the Zündapp KS 600 W, was also ordered by the Wehrmacht. The KS 600 W also went to the front in Wehrmacht gray, and was used to the war's end mainly as a heavy sidecar cycle.

The crew of this Zündapp K 800 W impressively demonstrates the use of this sidecar rig as a motorcycle rifle vehicle. The MG 34 could be fired while either standing or moving.

Below: Members of the bridge column of Armored Engineer Battalion 37 sat for a photo on their Zündapp KS 750 during service on the Greek Pelopennesus.

The "eyes right" of all the passengers suggests that these sidecar cycle crews are taking part in a parade. The cycles in the parade formation are Zündapps.

These young soldiers of an unidentified Luftwaffe intelligence unit sat on their Zündapp K 800 W for a souvenir photo in Italy in 1943-44.

This photo of a Zündapp KS 750 used by a Luftwaffe field division in 1943 shows that the heaviest sidecar cycles were made available to all Wehrmacht units after the motorcycle rifle troops were disbanded.

Motorcycle rifle machines in the Balkans in 1941. Some of the sidecars still have doors. In civilian motorcycle dealers' advertising brochures, these means of easy access were emphasized, but for military use they were more of a nuisance.

A big G on the sidecar of this Zündapp KS 600 W shows that it belongs to the Guderian Panzer Group. The drivers often had to cross such rough terrain (Russia, 1941).

Right: In 1935, Motorcycle Rifle Battalion 1 moved into its base at Langensalza. The unit was equipped with BMW R 11 cycles and had been formed from the former Cavalry Regiment 16 (mot).

This BMW R 11 sidecar cycle still saw service in Greece in 1941, though this model had been introduced into the Reichswehr in 1929. With 750 cc displacement, the R 11 was at that time BMW's most powerful sidecar cycle.

Although this Zündapp KS 600 W was equipped with a mount for an MG 34 on the sidecar for a gunner, it serves here as an officer's car after the air-landed operation on Crete in 1941.

According to Manual D 600, the Zündapp K 500 was not listed among the heavy sidecar cycles. This sidecar rig with IC (East Prussian) civilian plate was requisitioned by the Wehrmacht, and obviously was enjoyed by the children while in military service.

This photo is obviously from prewar times, since the vehicles have no headlight covers. The BMW R 12 rig, showing some dents in the sidecar, also makes a peacetime impression thanks to its good condition.

Everything on this BMW R 12 sidecar cycle that would burn has gone up in flames after the rig came under enemy fire. Notice the running boards for the driver and passenger, the folding part of the wide rear fender, and the unsprung rear wheel mount.

The rear fender of this BMW R 12 proved to be very unsuitable for off-road use. Mud stuck fast between the tire and the wide fender, and thus decreased the cycle's performance.

Right: The motorcycle crews on their unprotected cycles had to put up with bad weather conditions.

This Zündapp KS 750 was used as a B-Krad by a Luftwaffe field division in the Russian campaign in 1943.

Below: What was expected of sidecar cycles without off-road capability! The members of Panzerjäger Unit 7 demonstrate here that a BMW R 12 rig was quite capable of carrying five soldiers. The stable box frame of the R 12 could stand such a heavy load

What with the variety of motorcycles that were to be seen on all fronts during the course of the war, the numbers of very heavy Zündapp and BMW sidecar cycles must be regarded as modest. But whoever got the chance to drive one of these "war elephants" was probably as proud to be photographed as this messenger on his BMW R 75.

A BMW R 12 Wehrmacht cycle with modifications: The formerly wide front fender of the R 12 was later replaced by a narrow one that was less likely to fill up with mud in off-road use.

The BMW R 12 did not always come from military production, as shown by this picture of a formerly civilian R 12 sidecar rig, painted Wehrmacht gray and putting in service with the troops.

Below: At 580 kilograms, the K 800 W with sidecar was the heaviest model that Zündapp introduced before World War II. Here we see the sidecar cycle of a Luftwaffe unit in very well-kept prewar form.

Off paved roads, the cycles that were not off-road capable soon showed the limits of their usefulness. This was especially true in the mud season in the Russian campaign, during the spring thaw in the Caucasus region, where this Zündapp KS 600 W cycle got stuck in the muddy woodlands.

Any motorcycle that the industry had offered to the civilian sector as a socially acceptable means of transport in the mid-thirties was requisitioned when the war began, painted Wehrmacht gray, and turned over to the military. This BMW cycle with its bomb-shaped sidecar of "the best workmanship," as the manufacturer once advertised it, shared the same fate.

This BMW R 12 sidecar cycle of Long-range Reconnaissance Squadron 2 (F)./123 obviously ran out of fuel on the occupied British Isle of Jersey. But there was no lack of British cars to be requisitioned by this Luftwaffe unit in the late summer of 1940.

The crew of this BMW R 12 has painted the emblem of the "Grossdeutschland" Panzergrenadier Division on both cycle and sidecar. The picture shows clearly that the front fenders of these BMW cycles were made narrower yet again.

Despite driving for hours in rough, bumpy country, the motorcyclists and passengers (here an ostler) were usually able to look fresh and rested for a souvenir photo, as they do here on their BMW R 12. But reality was different; on their unprotected cycles the crews were exposed to wind and weather.

The driving school of this Luftwaffe unit had, among other vehicles, a BMW R 12 sidecar cycle. Training motorcycle drivers was of high value, and could be a matter of survival for the motorcyclists, who were often on their own in the last stages of the war.

These Zündapp KS 600 W cycles are being prepared for inspection.

German soldiers met their allies in the Balkans in 1941; their Zündapp KS 600 W shows the signs of their past war experiences.

These motorcycle riflemen have gathered around a BMW R 12 cycle for a souvenir photo. It is the sidecar cycle of a heavy machine-gun group of riflemen, as the machine-gun bracket on the sidecar shows.

Two BMW R 4 solo cycles made in 1932, and a BMW R 12 sidecar cycle lined up for a photo. While the solo motorcyclists were usually messengers and couriers, the drivers of the sidecar cycles often provided transport for such functionaries as doctors, ostlers, etc.

If necessary, motor vehicles, including these sidecar cycles, could be given a cleaning in shallow water. Surely this was not a pleasant job on a frosty March 21, 1941.

Members of a German paratroop unit are seen with their BMW R 75 in Italy in 1943-44. Obviously, this cycle had seen little action up to this point, as its external condition is very good.

The Wehrmacht's readiness for action with its supply of heavy sidecar cycles was limited when World War II began. Therefore, all obtainable civilian cycles were called in for military service, including this 1930 cycle, whose sidecar is mounted on simple leaf springs (France, 1940).

A training trip for members of Assault Gun Brigade 210 in its formative year of 1941, led by a Zündapp K 800 W sidecar rig. Note also the emblem of this unit—a tiger's head—applied to the headlight covers.

A Flak unit stops for servicing during the advance into Greece in May 1941. In the foreground is a BMW R 12 sidecar cycle carrying a spare fuel can instead of baggage packs on the sidecar.

A first-generation BMW R 12 with an attachment for an MG 34 on the sidecar. The deep cutout on the sidecar for the gunner is also easy to see. The sidecars of the previous model, the BMW R 11, usually were still equipped with a door.

The sidecar of this Zündapp K 800 W was modified to carry a light machine gun. The machine-gun mount, a later addition, is easy to see. The K 800 W had already proved to be a reliable off-road cycle before the war began, and was thus used by the motorcycle rifle troops since 1935.

The fact that the heaviest Zündapp and BMW sidecar cycles that were introduced by the Wehrmacht in the middle of World War II were suitable for use even on unpaved roads was attributable in part to their wide and knobby tires, as seen on this Zündapp KS 750.

Large numbers of BMW R 75 cycles in one place—as here—were seldom to be found.

This Zündapp K 800 W of a Luftwaffe unit has obviously been spared the rigors of service at the front, as the good condition of cycle and sidecar makes clear.

During the western campaign, many captured motorcycles were used by the German Wehrmacht. In the middle of this picture is a Belgian Rene Gillet with sidecar, with a BMW R 12 sidecar rig to the right of it.

The Zündapp KS 600 W cycle had already been supplied to the Wehrmacht in large numbers before the war began. This model can to be seen in many wartime photos, such as this one from the Russian campaign in 1941.

During the French campaign, many French cycles were captured and put to use by the Wehrmacht. This is a Gnome-Rhone X model made in 1937, its 18 hp and 800 cc classifying it among the Wehrmacht's heavy sidecar cycles.

Once this civilian Zündapp K 500 with its Steib sidecar must have stood in a salesroom, clean and shining, waiting for a buyer with sporting tastes. Here the cycle, covered with mud, is performing its service for a Luftwaffe field division in the Russian campaign.

Although the Zündapp KS 600 came much closer to meeting the Wehrmacht's needs than its K 500 predecessor, it was still not suitable for off-road use in wartime, all the more so with a sidecar. The cycle shown here is also shod with very narrow tires made for paved roads.

Who knows what made the members of Assault Gun Brigade 210 happy, but they seem glad to see this motorcyclist on his Zündapp KS 600 W at their camp.

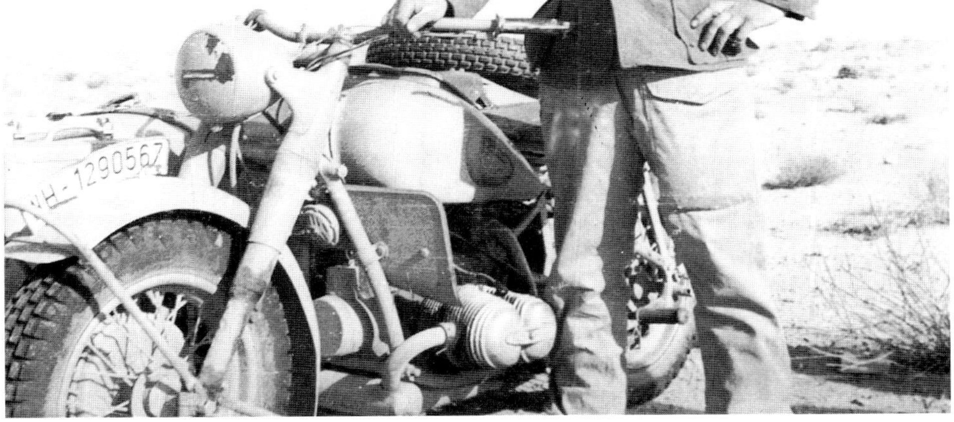

This BMW R 75 cycle was painted a sandy yellow, which indicates that it was used in North Africa. Yet this BMW sidecar cycle does not have a high air filter above the fuel tank, with which the R 75 was usually equipped for the North African theater of war. Notice, too, the great difference in the tires, which are not the slim, low-profiled type used on most Wehrmacht cycles.

This proud motorcyclist sits on his 1935 BMW R 12 machine, the most numerous, best-known, and best-loved BMW cycle in the Wehrmacht.

This scout or messenger platoon on DKW solo cycles is led by a requisitioned French Gnome et Rhone sidecar rig. A German service manual for this make and model was meant to help German motorcyclists maintain these cycles more easily.

Heavy sidecar cycles were also used by the military medical services. This BMW R 12, painted with a Red Cross emblem, was used by the medics in Italy in 1944.

Below: The Zündapp KS 750 sidecar cycle and the men of a Luftwaffe field division are splendidly turned out. The machine is equipped with a heating system that blew warm air at the driver and passenger. The crew could also enjoy quilted suits not available to every soldier in the Russian campaign.

No problems with engine cooling here! The driver of this BMW R 12 could barely see where the components of the engine were under the mud. Such conditions brought many a motorcycle's career to an untimely end.

Members of a motorcycle rifle battalion are ready to march on March 3, 1941, mounted on BMW R 12 sidecar cycles of the first type. The wide front fenders on some of the cycles are easy to see.

The crews of these Zündapp KS 600 W machines also belonged to a motorcycle rifle battalion.

At the end of the thirties, the Steib firm of Nürnberg advertised this type of sport sidecar for civilian use. Many of them, like the DKW 500, were requisitioned by the Wehrmacht for military use.

In the Russian winter war, many a motorcyclist had to begin his day by searching for his vehicle in the snow. The driver of this sidecar cycle still has a lot of work to do before he is ready for action (Russia, 1941-42).

What with the knobby profile of their tires, the heaviest Zündapp and BMW sidecar cycles handled off-road service better than other machines. This Zündapp KS 750 has just returned from a courier's mission through rough country.

An unidentified Panzerjäger company, uniformly equipped with Zündapp KS 600 W cycles, is seen maintaining its vehicles. In prewar days, motorcycle riflemen and other units that were supplied with cycles were always issued uniform models, making it easier to supply them with spare parts.

This DKW 500 NZ was subsequently fitted with a sidecar for wartime use. According to Manual 805/313 (Repair Shop Handbook) of May 2, 1944, the DKW SB 350 and SB 500 models, plus all NZ models, were suitable for attaching a sidecar.

This motorcycle messenger posed for his photo showing signs of service. His heavy sidecar cycle, an NSU 601 OSL, was introduced in the Wehrmacht in 1938 and 1939.

Neither the horse nor the sidecar cycle gives an impression of reliable service. The horse has surely gone without sufficient fodder of late, and the former civilian cycle, which saw service as the B-Krad of Assault Gun Brigade 666, was quite unable to handle the rough country.